PARROTS

by Mari Schuh

PEBBLE
a capstone imprint

Pebble Explore is published by Pebble, an imprint of Capstone.
1710 Roe Crest Drive
North Mankato, Minnesota 56003
www.capstonepub.com

Library of Congress Cataloging-in-Publication Data:
Names: Schuh, Mari C., 1975- author.
Title: Parrots / by Mari Schuh
Description: North Mankato, Minnesota : Pebble, [2020] | Series: Animals | Includes index. | Audience: Ages 7-8 | Audience: Grades 2-3 | Summary: "Parrots can do more than just mimic human speech. These colorful, sassy birds often mate for life, and some parrots can live 90 years or more. Learn more about these intelligent birds"– Provided by publisher
Identifiers: LCCN 2019037353 (print) | LCCN 2019037354 (ebook) | ISBN 9781977113443 (hardcover) | ISBN 9781977117977 (paperback) | ISBN 9781977113528 (adobe pdf) Subjects: LCSH: Parrots–Juvenile literature.
Classification: LCC QL696.P7 S355 2020 (print) | LCC QL696.P7 (ebook) | DDC 598.7/1–dc23
LC record available at https://lccn.loc.gov/2019037353
LC ebook record available at https://lccn.loc.gov/2019037354

Photo Credits
Shutterstock: A3pfamily, 23, BridgetSpencerPhoto, 9, Cuson, 18, dangdumrong, 24, David Roy Carson, 17, Dmitry Rukhlenko, 1, guentermanaus, 27, Herbert Eisengruber, 16, iweta0077, 25, Jiri Hrebicek, 13, Juriah Mosin, 5, Malgorzata Drewniak, 8, Nicole Kwiatkowski, 12, Petr Salinger, Cover, Phuong D. Nguyen, 20, Rafael Martos Martins, 7, Rosa Jay, 10, 11, SantiPhotoSS, 22, Super Prin, 14, Tracy Starr, 15, Wiratchai wansamngam, 21

Editorial Credits
Mandy Robbins, editor; Dina Her, designer; Morgan Walters, media researcher; Tori Abraham, production specialist

Dedication
For Logan and Ava—MS

Printed in the United States 4859

Table of Contents

Words in **bold** are in the glossary.

Amazing Parrots

It is a hot, wet day in the **rain forest**. A group of parrots gathers. They sit high in the trees. The parrots are loud and noisy! They call out. They make short chirps. They make loud squawks.

Parrots are colorful birds. They have curved beaks. There are more than 350 kinds of parrots. Macaws and parakeets are parrots. Lovebirds and lorikeets are parrots too.

Where Parrots Live

Most wild parrots live in the southern part of the world. They usually live in rain forests. These forests are warm or hot. Lots of rain falls there. Parrots can hide in the trees. They find lots of food in rain forests.

Parrots Range Map

Parrots live in other places too. Some live in deserts. These areas are dry. Some parrots live in snowy mountains. Others live in grasslands.

Parrots live in tree trunks. They find holes inside trees. They make nests there.

Some parrots live where there are no trees. They find other places to live. They take over **mounds** that termites build. They live inside cactus plants.

Some parrots live in cold areas. They make nests in the ground.

Parrot Bodies

Parrots are many sizes. Small ones are only 3.5 inches (9 centimeters) tall. Others are much bigger. They can be 40 inches (102 cm) tall.

Small parrots are light. Some weigh only 0.3 ounces (8.5 grams). Big ones are much heavier. Some big parrots can't even fly! These birds can weigh up to 6.6 pounds (3 kilograms).

Parrots are colorful. They have bright feathers. Many parrots have green feathers. They may also have red, yellow, and blue feathers. Other parrots are black, gray, or white. Many times, males and females look alike.

A parrot's bright colors blend in with the forest. It can be the same color as flowers and fruit. It can hide from danger.

A parrot has a short body. It has a big head and a short neck. Its beak is hooked. It is thick and strong. It can help the bird climb.

Strong feet help parrots hold onto branches. Parrots have four toes on each foot. Two toes point forward. Two point backward. Parrots use their feet to climb and grab food.

Eating

Parrots look for food in trees. They eat seeds and nuts. They also eat flower **buds**. Parrots find fruit to eat too.

Some parrots dig in the ground. They find bugs to eat.

A lorikeet visits flowers when it's hungry. It feeds on **pollen** and **nectar**. It has a special tongue. The tip is like a brush. It helps the bird sip nectar. The bird also sips juice from fruit.

Parrots use their feet and beaks to eat. They grab food with their beaks. The beaks are strong. They crack open hard nuts. Parrots can use their feet to lift food to their beaks.

Parrots taste food with their mouths. They can also taste with their throats.

What Parrots Do

Parrots are **social** birds. They live in groups called **flocks**. Some flocks have up to 30 birds. Living in a flock can help keep them safe. Some birds look for food. Others look out for danger.

Parrots look for food during the day. They can fly far to find food to eat. At night, the flock sleeps high in the trees.

Parrots are noisy. They make loud calls across the forest.

Parrots make noise for many reasons. They share information. They claim their home area, or territory. Sounds help them tell others where they are.

Parrots are smart birds. Pet parrots can copy what people say. They can learn many words.

Many parrots **mate** for life. Females lay two to six eggs at a time.

Chicks hatch after about 30 days. They are born nearly naked. They have their eyes closed. Both parents feed chicks. Chicks stay with their parents for three to eight weeks.

Parrots can live for a long time. Some live 60 to 80 years.

Dangers to Parrots

Many kinds of parrots are **endangered**, or in danger of dying out. They are in danger for many reasons. Trees where they live are sometimes cut down. People use the land to build houses and grow crops. They have fewer places to live. Animals such as monkeys and snakes eat parrot chicks.

Some people trap wild parrots. They sell the birds as pets. Parrots can die during the trip to the pet store.

Laws help keep many parrots safe. It is against the law in many places to trap and sell wild parrots.

Many people care about parrots. They work hard to keep them from dying out. They find ways to keep trees from being cut down.

Groups also make sure pet stores don't sell wild parrots. They want the birds to have long lives out in the wild.

Fast Facts

Name: parrot

Habitat: rain forests, woodlands, grasslands, mountains

Where in the World (range): southern half of the world

Food: seeds, fruit, nuts, nectar, pollen, and flower buds

Predators: hawks, eagles, snakes, monkeys

Life span: up to 80 years

Glossary

bud (BUD)—a small shoot on a plant that grows into a leaf or a flower

chick (CHIK)—a young bird

endangered (in-DAYN-juhrd)—in danger of dying out

flock (FLOK)—a group of the same kind of animal

mate (MATE)—to join with another to produce young

mound (MOUND)—a hill or pile

nectar (NEK-tur)—a sweet liquid found in many flowers

pollen (POL-uhn)—a powder made by flowers to help them create new seeds

predator (PRED-uh-tur)—an animal that hunts other animals for food

rain forest (RAYN FOR-ist)—a thick forest where rain falls almost every day

social (SOH-shuhl)—living in groups or packs

Read More

Bowman, Chris. *Parrots*. Minneapolis: Bellwether Media, 2015.

Dubke, Karon. *Pet Parrots Up Close*. North Mankato, MN.: Capstone Press, 2015.

Statts, Leo. *Parrots*. Minneapolis: Abdo Zoom, 2018.

Internet Sites

African Gray Parrot
www.enchantedlearning.com/subjects/birds/printouts/Grayparrotprintout.shtml

Amazing Animals: Name Calling
kids.nationalgeographic.com/explore/adventure_pass/amazing-animals/name-calling/

Parrots
easyscienceforkids.com/all-about-parrots/

Index